discovermore

A Look at Weather and Climate

Earth's Atmosphere

Marie Harts

Britannica Educational Publishing
IN ASSOCIATION WITH
THE ROSEN PUBLISHING GROUP

Published in 2025 by Britannica Educational Publishing (a trademark of Encyclopædia Britannica, Inc.) in association with The Rosen Publishing Group, Inc.
2544 Clinton Street, Buffalo, NY 14224

Distributed exclusively by Rosen Publishing.
To see additional Britannica Educational Publishing titles, go to rosenpublishing.com.

Editor: Brianna Propis
Book Design: Michael Flynn

Photo Credits: Cover Chaleephoto/Shutterstock.com; (series background) Dai Yim/Shutterstock.com; p. 4 Dima Zel/Shutterstock.com; p. 5 CHRISTIANTO/Shutterstock.com; p. 6 taka1022/Shutterstock.com; p. 7 AnnSky/Shutterstock.com; p. 8 Nomad_Soul/Shutterstock.com; p. 9 BlueRingMedia/Shutterstock.com; p. 10 Deni_Sugandi/Shutterstock.com; p. 11 mivod/Shutterstock.com; p. 12 Rainer Fuhrmann/Shutterstock.com; p. 13 Vasin Lee/Shutterstock.com; p. 14 Mtsaride/Shutterstock.com; p. 15 VectorMine/Shutterstock.com; p. 16 (left) Kay Cee Lens and Footages/Shutterstock.com; p. 16 (right) anmbph/Shutterstock.com; p. 17 David T. Cook/Shutterstock.com; p. 18 Elena Pimukova/Shutterstock.com; p. 19 ESB Professional/Shutterstock.com; p. 20 Pornprasert Khanchitchai/Shutterstock.com; p. 21 Leonid Sorokin/Shutterstock.com; p. 22 Tommy Daynjer/Shutterstock.com; p. 23 Toa55/Shutterstock.com; p. 25 (top) imageBROKER.com/Shutterstock.com; p. 25 (bottom) Diyana Dimitrova/Shutterstock.com; p. 26 Artsiom P/Shutterstock.com; p. 27 Elena11/Shutterstock.com; p. 28 BorneoRimbawan/Shutterstock.com; p. 29 Margaret.Wiktor/Shutterstock.com.

Cataloging-in-Publication Data

Names: Harts, Marie.
Title: Earth's atmosphere / Marie Harts.
Description: New York : Britannica Educational Publishing, in Association with Rosen Educational Services, 2025. | Series: Discover more! a look at weather and climate | Includes glossary and index.
Identifiers: ISBN 9781641901260 (library bound) | ISBN 9781641901253 (pbk) | ISBN 9781641901277 (ebook)
Subjects: LCSH: Atmosphere--Juvenile literature. | Earth (Planet)--Juvenile literature. | Weather--Juvenile literature. | Meteorology--Juvenile literature.
Classification: LCC QC863.5 W347 2025 | DDC 551.5--dc23

Manufactured in the United States of America

CPSIA Compliance Information: Batch #CSBRIT25. For further information contact Rosen Publishing at 1-800-237-9932.

Contents

Atmospheric Protection 4
Five Layers 6
A Variety of Gases 8
How Did the Atmosphere Form? 10
Weather or Not 12
Under Pressure 14
Water Vapor Creations 16
The Blue Sky Above 18
Greenhouse Gases 20
Burning Fossil Fuels 22
Making a Difference 24
Beyond Earth 26
Sustaining Human Life 28
Glossary 30
For More Information 31
Index 32

Atmospheric Protection

Earth, like many other planets, has an atmosphere. The atmosphere is a protective layer of gases surrounding the planet, like a shell.

Earth's atmosphere helps keep the planet warm enough for living things to survive. It protects life on the planet from being harmed by rays from the sun and other stars, as well as from incoming **meteoroids**.

Earth's atmosphere is made up of about 78 percent nitrogen, 21 percent oxygen, and 1 percent other gases.

If there was no atmosphere, Earth would become so cold at night that life could not survive on the planet.

Living things can exist on Earth because of the atmosphere. Not only does it keep us warm, but it holds the oxygen that humans and animals use to breathe. Certain gases in the atmosphere help plants survive too.

WORD WISE

A METEOROID IS A CHUNK OF ROCK OR METAL FROM SPACE. MOST METEOROIDS BURN UP AS THEY PASS THROUGH EARTH'S ATMOSPHERE BEFORE THEY CAN HIT THE PLANET, CAUSING A BRIGHT STREAK OF LIGHT CALLED A METEOR.

Five Layers

The atmosphere is divided into five separate layers. The layer closest to Earth is the troposphere. It extends up to 11 miles (18 km) at its highest point above Earth's surface.

The second layer is the stratosphere, which extends to about 30 miles (48 km) above the surface. The stratosphere contains the ozone layer. The ozone layer blocks some of the harmful rays that come from the sun. These rays can cause skin cancer, eye diseases, and other health problems in people and other animals.

The third layer is the mesosphere. It extends up to about 50 miles (80 km) above Earth's surface. The fourth layer, the thermosphere, ranges from about 50 to 375 miles (80 to 603 km) above Earth.

The clouds and flying birds you can see in the sky are located in the troposphere.

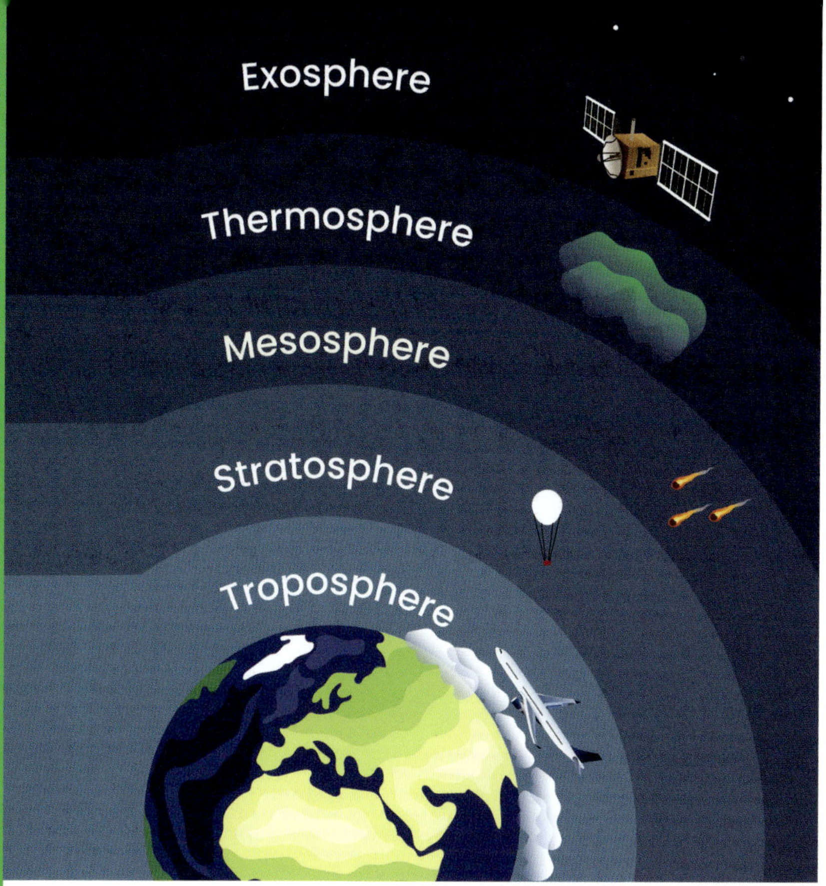

This illustration shows the layers of Earth's atmosphere, as well as what can be found in each layer.

The fifth and highest layer is the exosphere. It ranges from about 375 to 6,200 miles (603 to 9,977 km) above Earth. This layer ends where Earth's **gravity** is too weak to prevent particles of gas from drifting into space. It is where atoms and molecules escape into space and satellites orbit Earth.

WORD WISE

GRAVITY IS AN INVISIBLE FORCE THAT PULLS EVERYTHING DOWN TO THE GROUND ON EARTH'S SURFACE. THIS IS WHAT MAKES OBJECTS FALL AND ALSO WHAT GIVES OBJECTS WEIGHT.

A Variety of Gases

Several gases can be found in Earth's atmosphere, as well as small liquid and solid particles. The gases include nitrogen, oxygen, argon, carbon dioxide, helium, and water vapor.

Gravity pulls the heavier elements, such as nitrogen and oxygen, closer to Earth's surface. Lighter elements, including helium and hydrogen, are more common higher in the atmosphere.

In the lower regions of the atmosphere, weather patterns and wind evenly mix the gases. Elements are much less mixed in the higher regions.

Oxygen is needed for living things to breathe. Sometimes people use oxygen tanks if they have difficulty breathing or if they're deep-sea diving underwater, where oxygen is not found.

Gravity pulls objects toward the center of Earth. Mass is distributed equally in all directions surrounding the center.

compareandcontrast

When astronauts walk on the moon, their steps are bouncier than when people walk on Earth's ground. What does that say about the moon's gravity compared to Earth's?

How Did the Atmosphere Form?

It is still uncertain how the atmosphere formed. However, scientists believe it was the result of gases released from inside the planet and from tiny organisms, or living things.

More than 4 billion years ago, people would have been unable to live on Earth. The early atmosphere lacked the amount of oxygen that people need to survive. At that time, the only organisms on Earth were anaerobic. This means they could survive without oxygen.

The atmosphere was most likely formed from gases spewed, or rapidly released, from volcanoes.

Organisms like cyanobacteria, or blue-green algae, added oxygen to the atmosphere as it was forming.

In time, some organisms developed photosynthesis. Photosynthesis is a process that allows organisms, such as plants, to produce their own food. Through photosynthesis, organisms take in energy from sunlight. They use this light energy to change water and a gas called carbon dioxide into oxygen and nutrients called sugars. The oxygen is released into the air and is built up in the atmosphere over time.

Consider This

The oxygen people need comes from trees and plants. Why is it important to care for the environment? What would happen if trees and plants began to quickly die out?

Weather or Not

Weather is the daily state of the atmosphere, such as being hot or cold, wet or dry, and calm or stormy. Earth's weather occurs mainly within the troposphere. Weather is important because it affects people's comfort, food supply, and safety.

Wind can be a gentle breeze or a large gust that knocks down traffic lights!

Lightning is electricity released from Earth's atmosphere during thunderstorms. Thunder is created when lightning passes through the air.

There are five features in the atmosphere that combine to create weather. The first is temperature, or how hot or cold it is outside. Wind, the movement of air across Earth's surface, is the second feature. The third feature, humidity, is the amount of moisture in the air. Precipitation, such as rain, snow, and sleet, is the fourth feature. The last feature, atmospheric pressure, is the weight of the air.

Scientists who study weather are called meteorologists. They pay attention to the features in the atmosphere in order to predict what the weather will be like in a few hours or a few days. When changes in atmospheric pressure occur, thunderstorms can happen.

Consider This

Why is it helpful to be able to know an upcoming day's or week's weather ahead of time?

Under Pressure

Atmospheric pressure—also known as barometric pressure—is the weight of the air. The gases in the air are made of particles called molecules. The molecules are pulled toward Earth by gravity. The pressure is therefore greatest at sea level. There, the gas particles are pressed together by the weight of the air above them.

Air becomes lighter farther away from Earth's surface. This is because the air molecules are always moving. Those farthest away from the surface have more room to move. When there is less pressure on them, they are able to spread out more.

Barometric pressure is measured with a barometer. Barometers have been used since the 1600s.

ATMOSPHERIC PRESSURE

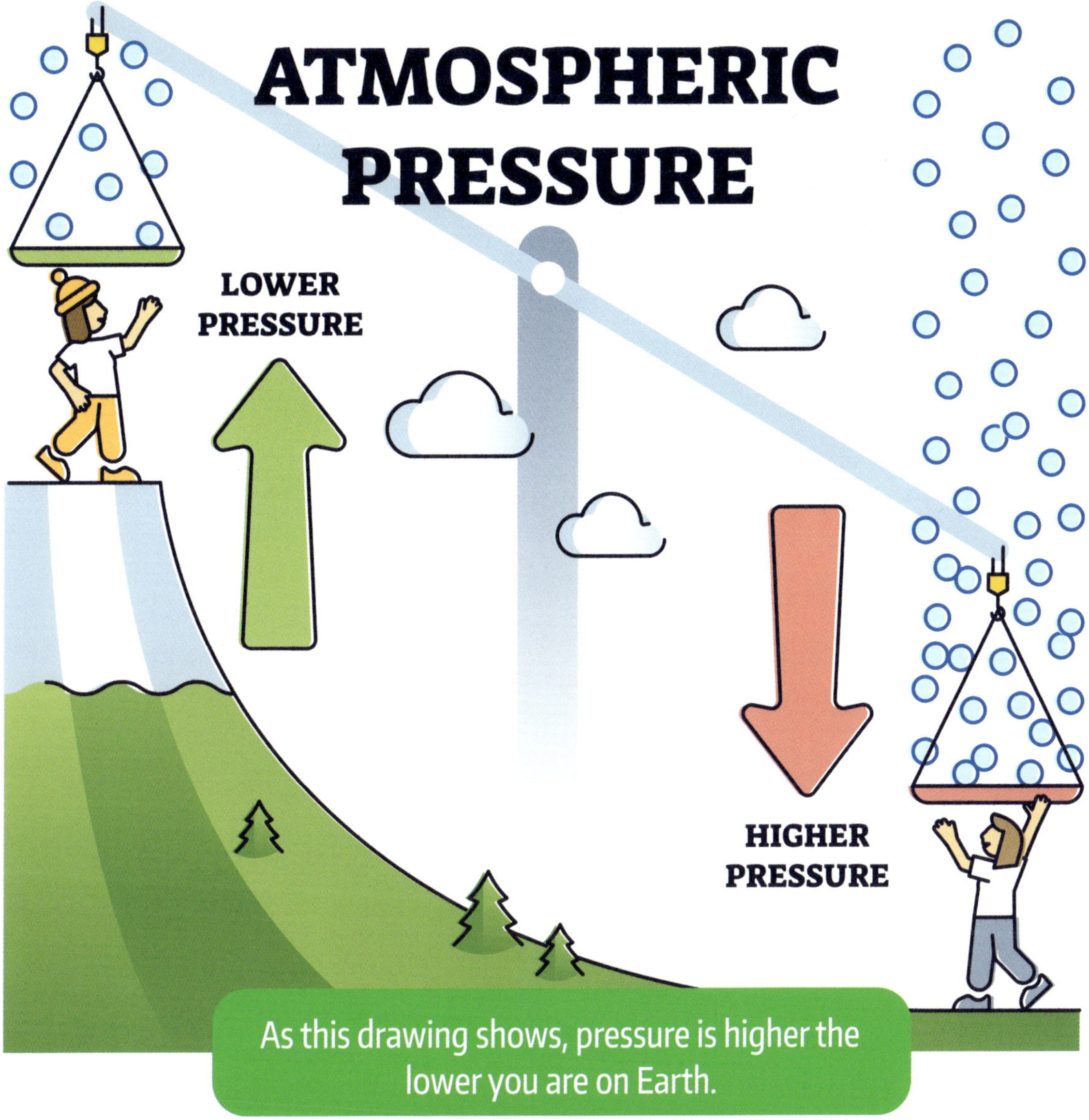

As this drawing shows, pressure is higher the lower you are on Earth.

compare and contrast

Why is it more difficult for people to breathe on tall mountains than at sea level?

Water Vapor Creations

When water is heated, it evaporates, or turns into a gas. This happens to bodies of water on Earth, such as oceans, lakes, rivers, streams, and ponds. That gas is called water vapor. It rises into the air. The amount of water vapor that air can hold depends on the air's temperature. The cooler the air, the less water it can hold. As air cools, some of the water vapor condenses, or turns back into a liquid. It forms visible water droplets. When the droplets come together they form clouds. Once formed, clouds remain until the air warms or until the droplets become so large and heavy that they fall to the ground as precipitation.

Can you identify the types of clouds in these photos?

A special type of cloud that combines the three forms is called a nimbus cloud. Most precipitation falls from this kind of cloud.

Meteorologists identify clouds mainly by appearance. The three main types of clouds are cirrus, cumulus, and stratus. Cirrus clouds are wispy, feathery, and made entirely of ice crystals. Cumulus clouds are shaped like fluffy cotton balls and are usually in the sky when the weather is nice. Stratus clouds are low-hanging and layered.

Consider This

Different types of clouds bring different types of precipitation. Why do you think this is?

The Blue Sky Above

The white light from the sky above is from the sun's rays streaming down to Earth. White light is made up of all the colors of the spectrum, and each color has a different **wavelength**. For instance, red has a longer wavelength than blue. As the sun's rays pass through the atmosphere, they are reflected and bent by air molecules and dust particles.

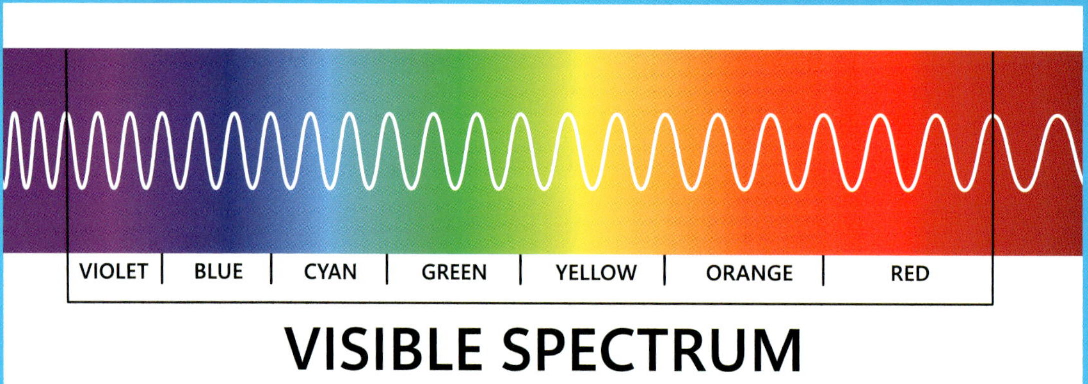

Colors look different from one another because they each have different wavelengths.

If the wavelength of the color blue was longer, the sky would be a different color.

The short blue light waves get absorbed by particles in the atmosphere. The blue light gets scattered, or bounced from particle to particle, in a process called diffusion. This is why the sky above appears to be blue.

WORD WISE

WAVELENGTH IS THE DISTANCE BETWEEN TWO HIGH POINTS OF A WAVE. LIGHT TRAVELS IN THE FORM OF A WAVE.

Greenhouse Gases

The greenhouse effect occurs when heat is trapped near Earth's surface from gases in the atmosphere. These gases are called greenhouse gases. The most common greenhouse gases are water vapor, carbon dioxide, and methane.

Land, oceans, and plants absorb, or soak up, energy from sunlight. They release some of this energy as heat. Greenhouse gases absorb the heat and then send it back toward Earth. Without greenhouse gases, this heat would escape back into space.

Greenhouse gases can contribute to respiratory disease in people through air pollution and smog.

In 2022, emissions from human activity grew by 0.9 percent, or 321 million tons (291 million mt).

Scientists have discovered that human activities are increasing the power of the greenhouse effect. When people drive cars or operate factories, they burn coal, oil, and other fossil fuels. This adds extra greenhouse gases to the air, and the extra gases trap more heat. Scientists believe this is a leading cause of global warming, or an increasing climb in the temperature of Earth's surface.

Consider This

How can people reduce their release of greenhouse gases into the atmosphere?

Burning Fossil Fuels

When harmful substances are released into the air, pollution happens. Wildfires, volcanoes, and chemicals from factories cause some air pollution. But most air pollution comes from burning fossil fuels. These include coal, oil, and natural gas. Factories, electrical plants, and automobiles use these fuels for power. When the fossil fuels are burned they release solid particles, such as ash and soot, into the air. Harmful gases may also be released. This type of pollution is often seen as smog over cities. Smog-filled air is hazy, or hard to see through.

One of the animals impacted by global climate change is the sea turtle. Its nesting areas are vulnerable to sea-level rise, increased storminess, and changing temperatures, which are all impacts of climate change.

Cars that run on fossil fuels release toxins into the air.

In addition, some kinds of air pollution cause global warming. Scientists have discovered that the average surface temperature on Earth is slowly rising. Higher temperatures could cause polar ice caps to melt. This would cause sea levels to rise. Plants, animals, and buildings along coastlines would be in danger. Some forms of air pollution are also harmful to the ozone layer in the stratosphere. The ozone layer makes life on Earth possible because it absorbs dangerous rays from the sun.

Consider This

Think of the machines and technology in your home and around your neighborhood. Which of them burn fossil fuels?

Making a Difference

Global warming and air pollution can have harmful impacts on the future. The rising temperatures may lead to changes in rain patterns. This could bring about flooding in some areas and drought in others. Both drought and flooding could hurt farmers. Drought can be a problem in forests because it increases the danger of forest fires. Coastal flooding caused by rising sea levels may force many people to move farther inland.

Many scientists and governments, however, are working to reduce the damage people have caused. Some governments have passed laws that require factories and companies to reduce the amount of pollutants they put into the air. Some companies and individuals are working to develop new ways to use other sources of energy, such as solar power and wind-generated electricity. These sources do not pollute as much as burning fossil fuels does.

There are ways for people to individually reduce their fossil fuel emissions.

Rising temperatures have caused large glaciers, or ice sheets, to melt. This harms the animals, such as polar bears, that live in Arctic areas.

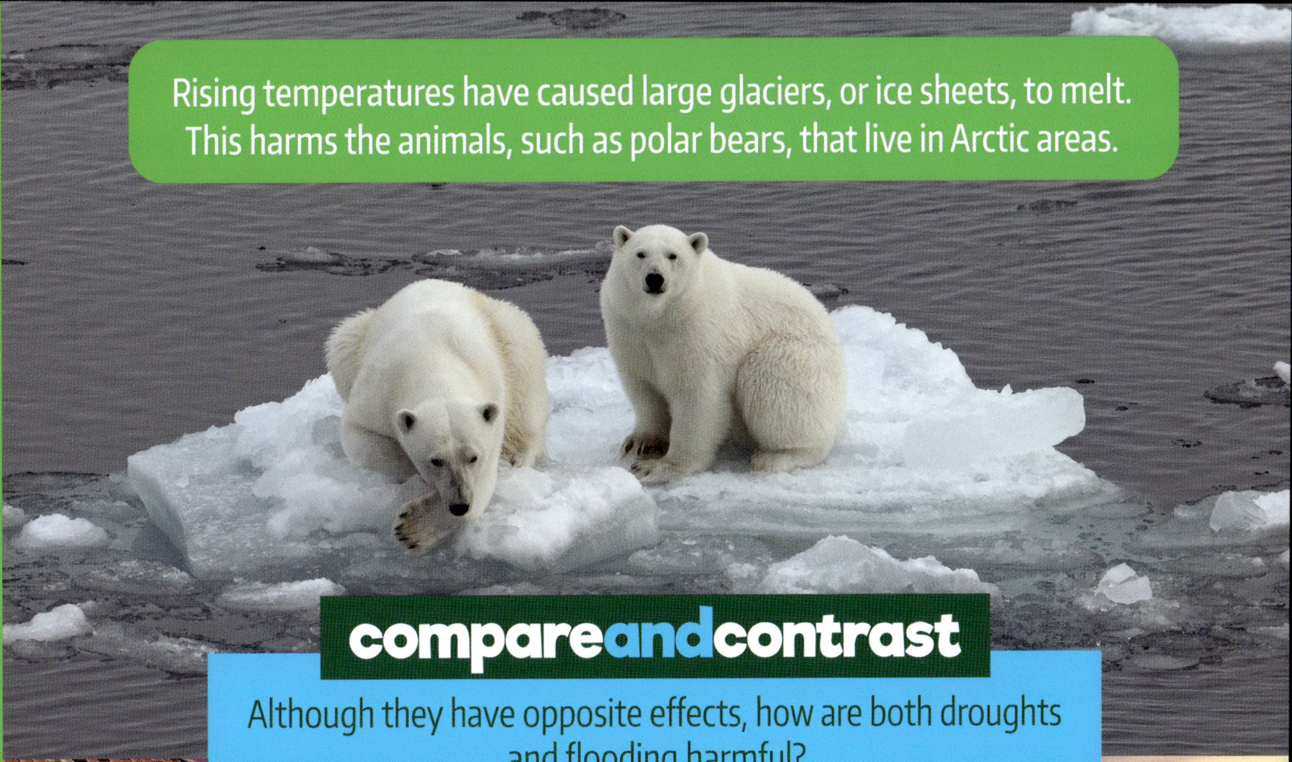

compare and contrast

Although they have opposite effects, how are both droughts and flooding harmful?

Solar-powered technology is an alternative to burning fossil fuels. Solar electricity, solar lighting, and solar heating are a few examples of things that are powered by converting energy from the sun.

Beyond Earth

Earth's atmosphere is unique because it can support living things, such as animals and plants. It protects people from the sun's harmful rays. It contains the right amounts of gases that are needed to breathe. It prevents Earth from becoming too cold or too hot.

The clouds in Venus's atmosphere are thick and hide the harsh surface below. They are also full of acidic droplets.

Saturn's atmosphere has winds that reach speeds of 1,118 miles (1,800 km) per hour.

There are 8 planets and more than 160 moons in our **solar system**, so there are many different kinds of atmospheres. For instance, Venus has a thick atmosphere. Its surface temperature is about 867°F (464°C), making it too hot for life as we know it to exist. In contrast, the atmosphere on Mars is very thin. Carbon dioxide makes up about 95 percent of the atmosphere of Mars—meaning it would be impossible for animals to breathe on Mars.

WORD WISE

THE SOLAR SYSTEM IS THE COLLECTION OF THE 8 PLANETS AND THEIR MOONS IN ORBIT OF, OR TRAVELING AROUND, THE SUN. IT ALSO INCLUDES THE ORBITS OF ASTEROIDS, METEOROIDS, AND COMETS.

Sustaining Human Life

Earth is one of an endless number of objects in the universe, yet scientists still have not discovered any other place like it. Its unique atmosphere is what allows life to thrive. It makes it possible to sustain, or support, human life—from children playing in the snow to farmers growing crops.

Earth's atmosphere protects the planet's many environments, from rainforests to the snowy Arctic.

In order to continue having Earth as our home, we must protect the planet by being careful about releasing harmful toxins into the air.

Scientists have discovered many planets and moons that have atmospheres. Some have the same gases that Earth has, including water vapor. Scientists have even seen lightning on Jupiter and Saturn. No other atmosphere, however, comes close to matching Earth's. Earth would be just another object spinning in space if it wasn't able to sustain living things.

Consider This

What do you think are the most important things Earth's atmosphere does to protect the planet and those who live on it?

Glossary

acidic Containing acid, a powerful chemical.
asteroid A small rocky body that orbits the sun.
comet A small, icy solar system body.
emission Something produced and released, such as harmful substances.
food supply Available food production for a population.
greenhouse effect Warming of Earth's surface and part of the atmosphere caused by gases in the air that trap heat.
mass The amount of matter in something.
polar ice cap A sheet of ice that forms around the North and South Poles.
ray A stream of energy that enters the atmosphere from outer space.
smog Fog or haze that is combined with air pollutants.
toxin A poisonous substance.

For More Information

Books

Jankeliowitch, Anne. *Solar System*. Cambridge, MA: Barefoot Books, 2019.

Meinking, Mary. *Clouds*. Minneapolis, MN: Cody Koala, 2019.

Shofner, Melissa Rae. *Earth's Atmosphere*. Buffalo, NY: PowerKids Press, 2019.

Websites

Atmosphere – National Geographic Society
education.nationalgeographic.org/resource/atmosphere/
Read more about the layers of Earth's atmosphere.

Solar System Exploration – NASA
science.nasa.gov/solar-system/
Explore the solar system and the countless bodies in the universe.

Ten Basic Clouds – NOAA
www.noaa.gov/jetstream/clouds/ten-basic-clouds
Learn more about the types of clouds in the sky above.

Publisher's note to educators and parents: Our editors have carefully reviewed these websites to ensure that they are suitable for students. Many websites change frequently, however, and we cannot guarantee that a site's future contents will continue to meet our high standards of quality and educational value. Be advised that students should be closely supervised whenever they access the internet.

Index

A

atmospheric pressure, 13, 14–15

C

carbon dioxide, 8, 11, 20, 27
clouds, 6, 16–17
colors, 18–19

E

exosphere, 7

F

fossil fuels, 21, 22, 23, 24

G

global warming, 21, 23, 24
gravity, 7, 8, 9, 14
greenhouse gases, 20–21

M

mesosphere, 6, 7
meteoroids, 4, 5, 27
meteorologist, 13, 17

N

nitrogen, 4, 8

O

oxygen, 4, 5, 8, 10, 11
ozone layer, 6, 23

P

photosynthesis, 11
pollution, 22, 23, 24
precipitation, 13, 16, 17, 24

S

solar power, 25
solar system, 26–27, 29
stratosphere, 6, 7, 23
sun, 4, 6, 11, 18, 20, 23, 25, 26, 27

T

temperature, 13, 16, 24
thermosphere, 6, 7
troposphere, 6, 7, 12

V

volcanoes, 10, 22

W

water vapor, 8, 16, 20, 29
wavelength, 18–19
weather, 8, 12–13